ENGINEERING ESSENTIALS FOR STEM INSTRUCTION

How do I infuse real-world problem solving into science, technology, and math?

Pamela
TRUESDELL

ASCD | NSTApress

Website: www.ascd.org
E-mail: books@ascd.org

Website: www.nsta.org
E-mail: nstapresseditorialof@nsta.org

Published simultaneously by ASCD, 1703 N. Beauregard Street, Alexandria, VA 22311 (www.ascd.org) and National Science Teachers Association, 1840 Wilson Blvd., Arlington, VA 22201 (www.nsta.org). ASCD is a nonprofit, nonpartisan membership association that provides expert and innovative solutions in professional development, capacity building, and educational leadership to educators, schools and school districts; NSTA is a nonprofit, nonpartisan membership association for science teachers from kindergarten through college.

PAPERBACK ISBN: 978-1-4166-1905-5   ASCD product #SF114048

Also available as an e-book (see Books in Print for the ISBNs).

Library of Congress Cataloging-in-Publication Data
Truesdell, Pamela.
  Engineering essentials for STEM instruction : how do I infuse real-world problem solving into science, technology, and math? / Pamela Truesdell.
     pages cm.—(ASCD Arias)
  Includes bibliographical references.
  ISBN 978-1-4166-1905-5 (paperback : alk. paper) 1. Science—Study and teaching (Higher) 2. Technical education. 3. Engineering—Study and teaching (Higher) 4. Mathematics—Study and teaching (Higher) I. Title.
  Q181.T84 2014
  620.0071'1—dc23

                                                                    2014005291

22 21 20 19 18 17 16 15 14   1 2 3 4 5 6 7 8 9 10

ENGINEERING ESSENTIALS FOR STEM INSTRUCTION

How do I infuse real-world problem solving into science, technology, and math?

Want to earn a free ASCD Arias e-book?
Your opinion counts! Please take 2–3 minutes to give
us your feedback on this publication. All survey
respondents will be entered into a drawing to
win an ASCD Arias e-book.

Please visit
www.ascd.org/ariasfeedback

Thank you!

Engineering and Its Role in STEM

Many educators today say that the traditional approach to teaching science, technology, engineering, and mathematics is outdated, and that the STEM subjects should be taught together rather than as totally separate, "siloed" disciplines. In this publication, we will look at how engineering, the "E" in STEM, can unify all four subject areas.

The best and most engaging way to achieve this unity is through engineering projects that ask students to design solutions for real-world problems. Consider that in the mathematics classroom, problem solving has long been promoted as the way for teachers and students to climb up the Bloom's taxonomy pyramid. Engineering projects steer students past simple questions of how many apples Sally has and toward authentic problem-solving situations. Whereas students in science class can sometimes get bogged down in following a series of steps to verify an accepted scientific fact, engineering projects open their eyes to the discipline's true nature. Projects that ask students to apply current knowledge and exploration to new areas in pursuit of the elusive "best" solution make them active players in the world of science. Finally, when students see technology through the lens of engineering, they understand that it's much more than a synonym for "something that can be plugged in." Engineering both drives innovation of technology and uses technology to create advancements in the world around us.

What Is Engineering?

Before we consider how to introduce engineering into science, math, and technology instruction, there are some questions to be answered. Perhaps the most basic are "What is engineering?" and "What do engineers really do?"

Defining engineering and the work of engineers is somewhat like defining medicine and the work of those who practice it. In the field of medicine, there are surgeons, doctors, nurses, researchers, technicians, and many other kinds of workers. The challenge of coming up with a definition of what all these people do is compounded by the field's many branches—cardiology, dermatology, pediatrics, psychiatry, and so on. Still, it's a bit easier for us to grasp the "whole" of medicine because the average person actually comes in contact with doctors and nurses. We have the opportunity to talk to our X-ray technician, surgeon, pharmacist, or physical therapist. By comparison, very few of us will meet the civil engineer who designed the bridge we drive across on the way to work, or the chemical engineer who came up with the laundry detergent formula that makes our whites whiter.

A full examination of engineering and engineering education would require many more words than this format allows. But in the same way that a tourist heading to Paris can find in a Michelin guidebook ample information to make a stay in the City of Light more rewarding, in this publication's overview of engineering, you'll find guidance to help you bring engineering into the classroom in a more meaningful way—and make a real, positive difference in your students' learning.

Let's begin our tour by looking at how engineering is generally defined. According to the *American Heritage Dictionary* (2009), *engineering* is "the application of scientific and mathematical principles to practical ends such as the design, manufacture, and operation of efficient and economical structures, machines, processes, and systems." Most general dictionaries define engineering similarly—as an application of math and science.

When we turn to the definitions offered by engineering-based groups, there's a definite shift toward a more specialized meaning. The American Society for Engineering Education produces a website and publication called *Engineering, Go For It*, or *eGFI* for short. (Available at www.egfi-k12.org, it is a great source for teachers and students interested in learning more about engineering fields.) In the October 2009 issue, *eGFI* states the following:

> Engineers solve problems using science and math, harnessing the forces and materials in nature. They draw on their creative powers to come up with quicker, better, and less expensive ways to do the things that need to be done. And they find ways to make dreams a reality. (p. 2)

The difference from the standard dictionary definition is subtle but important. In the *eGFI* definition, and for engineers, *solving problems* comes first. However, we want to be careful not to translate this statement into something oversimplified, such as "to solve problems in math or science classrooms is to do engineering." After all, at one time it was popular to try to teach critical thinking skills by giving students word problems or story problems. The trouble was, these problems were often one-dimensional and had little relation to the world in which the students lived. Engineers derive the problems they tackle from the real world around them. Yes, in the classroom, for educational purposes, it is sometimes necessary that problems be "made up" rather than actual, but they should never be simplistic or irrelevant.

Tools for Engineering

Having established that solving problems is the main goal in engineering, the next question to consider is how engineers go about that work. As the statement in *eGFI* indicates, they make use of math and science. In fairness to the "T" in STEM, they also make use of technology. But math, science, and technology are only some of the tools that engineers use. Depending on the field of engineering, an engineer's toolbox can be filled with a tremendous assortment of methods to solve whatever problem is on the table. Let's look at the most common tools at an engineer's disposal.

Engineers generally have and need inherent *curiosity*. When I worked in a public engineering high school that had no entrance requirements, one of our principals advocated evaluating prospective students' suitability for the program by asking them if they were good in math and science. I told the principal that it would be better to ask students if they liked to take things apart to see how they work. I knew that if students had inherent curiosity, we could teach them the necessary math and science. They would learn those subjects because they would see the need for them. On the other hand, it is quite difficult—although not impossible—to teach an incurious child to be curious.

Creativity is another essential tool, and another that is easier to support and channel than to develop from scratch. Even though some fields of engineering seem to consist of just following rules and regulations, there is always creativity below the surface. For example, civil engineers need to understand building codes, material strengths, and framing techniques. However, these engineers must also figure out ways to bring older buildings up to code, use new products, and satisfy clients who want something done that has never been done before. Other fields, such as computer engineering, are more obviously prone to divergent thinking. It may be a cliché, but engineers truly have to be able to "think out of the box," or new and innovative products won't happen.

Organization and *logic* may appear to be the opposite of creativity, but it is important for engineers to have these tools, too. As a part of my fellowship at the National Science Foundation, I visited 27 different universities that had grants for the Research Experiences for Teachers (RET) program. In many cases, I was able to interview teachers while they were involved in the program. Ryan Cain, a teacher at NYU-Poly's RET, talked about his interest in the maker/tinker movement. He said that he is the type of science teacher who loves to play with things to figure them out. When working on transparent soils in the RET's engineering research lab, his "reaction was to act as a tinkerer, but [the program leaders] made sure we acted as engineers." Due to time constraints, material costs, and other considerations, engineers must be logical in their problem-solving approach.

Likewise, in order to operate within the constraints of a given project, engineers must be able to use *clear and concise problem formulation.* Problems must be stated very specifically. "Find a way to provide clean water for third world countries" is too broad a challenge to tackle. "Create a filtration system using relatively accessible materials that is cheap enough to be deployed at scale in third world countries" may still be challenging, but it is much more focused. This type of focus is also needed in *decision making.* Engineers use critical thinking skills to consider the constraints, weigh the possible solutions, and come to an objective conclusion.

Because the *American Heritage Dictionary* is not an authority to be undervalued, *knowledge of mathematics and science* is unquestionably an essential tool for engineering, as is *technical knowledge* specific to various particular fields of engineering. This is not just "head knowledge" but practical knowledge as well. An engineer must be able to use mathematical and scientific understanding to analyze a problem, describe it, talk about it, and successfully find and execute a solution.

Many people have images in their heads of solitary engineers examining blueprints or diagrams—perhaps hunched over an old-school drawing board or peering at a computer-aided drafting (CAD) image on a screen. However, much of what engineers do relies on the *ability to work in teams*. When facing a complicated problem, having a group of people (each of whom brings unique experiences and expertise) work together is often a more productive approach. *Communication skill* is also essential when dealing with clients; it's what allows engineers to understand client needs and make sure that their needs are met.

Not every challenge an engineer faces requires the use of all the aforementioned tools. In fact, maybe the most valuable tool in the engineering toolbox is one we have not yet discussed—the one that helps engineers choose the right tools for each job. It is the engineering design process.

The Engineering Design Process

Most people try to avoid problems, but bringing a problem to a solution is what engineers live for. In order to solve problems, engineers follow what is called the *engineering design process*—the EDP. It is as important to engineering as what's popularly known as "the scientific method" is to science and the various mathematical algorithms are to mathematics. Just as with scientific processes and mathematical algorithms, the EDP was not carved into stone tablets and handed to Leonardo DaVinci on top of a mountain. It comes in various versions that are known by different names, but in all its forms, the essential steps remain the same. I'd like to look at the two versions of the EDP that are the dominant players in the pre-college educational arena.

Starting at end of the spectrum aimed at younger students, Engineering is Elementary (EiE), a program developed by the Museum of Science in Boston, was designed to "take advantage of the natural curiosity of all children to cultivate their understanding and problem-solving in engineering and technology" (Cunningham, 2009, p. 11). In order to present the EDP in a format that's easy for the young target audience to understand, EiE uses a simple, five-step process with a single key word describing each step (see Figure 1).

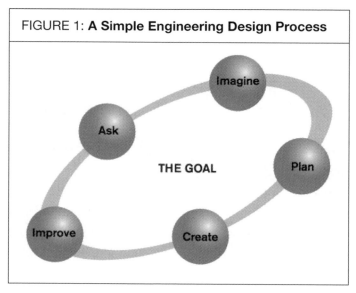

FIGURE 1: **A Simple Engineering Design Process**

Source: Engineering Is Elementary (2014).

At the other end of the age spectrum is the version of the EDP articulated by the International Technology and

Engineering Educators Association (ITEEA) and long used by Project Lead The Way (PLTW). Started in 1997 by Richard Blais, a high school technology teacher in upstate New York, PLTW's STEM curriculum is now in more than 5,000 schools in all 50 U.S. states and the District of Columbia. ITEEA's model presents the EDP in 12 steps (see Figure 2).

FIGURE 2: A 12-Step Engineering Design Process

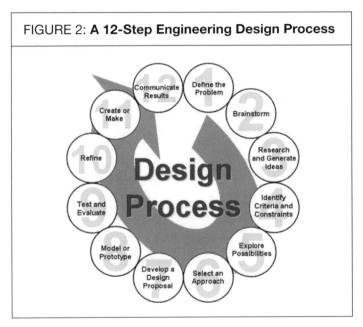

Source: "Design Process Overview" from *Introduction to Engineering Design, 2007–2008*. Copyright 2008 by Project Lead The Way, Inc. Used with permission.

Let's use this detailed approach to examine the engineering design process more closely.

1. Define the Problem. Problems (needs or wants) can be presented by a client or identified by the engineer. It is important to be specific.

2. Brainstorm. Teamwork is crucial for this step; presenting ideas in a group creates a synergy that cannot be generated by individuals working alone. The goal is to seek a quantity of ideas and not worry about the quality. Ideas should not be praised or dismissed. Sometimes the most outlandish or crazy ideas can be the source of a groundbreaking product or invention.

3. Research and Generate Ideas. Research any existing solutions and identify why they are not adequate or appropriate. Investigate who is affected by the problem, and consider the problem from their perspectives (What do they need? What do they want?). Interview the people affected, if possible. Perform market research to determine the degree to which a want or need exists.

4. Identify Criteria and Specify Constraints. Redefine the problem in a way that both the client and the engineering team can agree on. Determine what the solution must do. Identify the constraints, including time and budget. Compile the information into a *design brief*—a written plan that identifies the problem to be solved, its criteria, and its constraints. The purpose of the design brief is both to articulate as many aspects of the problem as possible before attempting a solution and to make sure that all parties are on the same page.

5. Explore Possibilities. Brainstorm new solution ideas in light of the design brief. Come up with alternatives to existing products, based on new materials and methods. Consider brand-new and unique approaches.

6. Select an Approach. Revisit previous steps and answer any remaining questions. Come to a final team consensus or compromise by voting or creating a *decision matrix*. Also known as a Pugh chart, a decision matrix lays out the attributes of the various ideas so that they can be considered in an objective way. For an example, see http://bit.ly/LU1b18.

7. Develop a Design Proposal. Create detailed and annotated sketches. Decide on the type(s) of material and manufacturing methods. Make computer models. Generate working drawings from the computer model(s) so that the idea can be built.

8. Make a Model or Prototype. Make models (scaled models or mock-ups) to study shape, fit, or texture. Build a full-size, functional prototype for testing purposes.

9. Test and Evaluate the Design Using Specifications. Test the prototype in controlled and actual conditions. Gather performance data. Evaluate results to identify possible concerns or shortcomings. Determine if redesign work is needed.

10. Refine the Design. Reassess the validity of the design criteria and make adjustments to the design brief, if necessary. Work back through the design process until the solution satisfies the design criteria. Update the documentation of the final solution.

11. Create or Make the Solution. Determine how the item will be produced, including what type of packaging will be needed, if any.

12. Communicate Processes and Results. Present the engineering team's final solution to the initial problem through presentations with visual aids (e.g., slide show, models, prototype). Develop written reports with appropriate graphic documentation (e.g., charts, graphs, technical drawings). Market and sell the product.

Even though the steps are numerically ordered, PLTW emphasizes that engineers can skip steps or jump back to a previous step, depending on the challenge, the constraints, and other variables. This idea of malleability is also conveyed in the EiE take on the engineering design process. There, the graphic illustration of the approach omits numbers and does not designate the beginning or the end of the cycle. Whichever specific sequence the EDP takes, it always revolves around accomplishing of the goal and solving the problem.

In these two systems, and in any version of the EDP, there are some very obvious similarities. First, the process is circular. One of the keys of engineering is that it is an iterative process. The Next Generation Science Standards (NGSS) refer to this process as "optimization." When an engineer "solves" a problem, the process is not over. Any problem always has another solution lurking out there that is cheaper, faster, or easier to produce. Given the advances in technology (e.g., polymers, integrated circuits), the key to those new solutions may not have existed when the problem first hit the drawing board. It is crucial that students grasp the iterative aspect of engineering.

The Engineering Design Process, Illustrated

Because the EDP is so crucial to understanding and using engineering in the classroom, before moving on, let's take a moment to go through an illustration.

When I taught a 9th grade engineering course, one of my favorite videos to show was an episode of ABC News' *Nightline* (1999) illustrating how a design firm (IDEO, based in Palo Alto, California) met the challenge to redesign the familiar grocery store shopping cart. This video is available on YouTube, and I highly recommend watching it and sharing it with students. Here's how the IDEO engineers' approach maps to the engineering design process we have outlined.

1. Define the Problem. ABC News (the client) asks IDEO, an industrial design firm, to redesign the shopping cart to bring it into the 21st century. The challenge, as stated in the video, is to "take something old and familiar, like the shopping cart, and completely redesign it in just five days."

2. Brainstorm. The team from IDEO enters the "Deep Dive"—full immersion in the problem. The video explains that IDEO's guide to brainstorming includes principles like "one conversation at a time," "stay focused on the topic," "encourage wild ideas," "defer judgment," and "build on the ideas of others." The team's proposed solutions, some of which include drawings, are put up on the wall.

3. Research and Generate Ideas. In their own discussions, members of the IDEO team determine that cart safety and cart theft seem to be important issues. The team goes out to interview the "experts"—grocery store staff and shoppers. While on site at a grocery store, they observe and photograph people's habits and shopping methods. Through this work, the team discovers some important information: that there are 22,000 reported cases per year of children requiring hospitalization for injuries sustained from shopping carts; that plastic carts have greater surface area, which is problematic for unattended carts that are outdoors during high winds (one cart was clocked rolling across a parking lot at 35 miles per hour); and that "professional" shoppers leave carts at strategic locations in the store, gather their items, and bring them back to their carts.

4. Identify Criteria and Specify Constraints. It is not shown in the *Nightline* episode, but the design brief for this project would look something like this:

- Client Company: ABC News
- Target Consumer: Grocery store owners
- Designer: IDEO
- The Problem: The old and familiar shopping cart is inefficient, unsafe, prone to theft, and unattractive.
- The Challenge: Design, prototype, and test an attractive shopping cart that will deter theft and make the supermarket experience easier, safer, and more efficient.
- Criteria and Constraints: A five-day deadline, and the design must "nest" for storage, as traditional shopping carts do.

5. Explore Possibilities. The team decides to explore four different ideas, each addressing one of the four areas of concern: ease of shopping, safety, fast checkout, and the ability to find what you are looking for.

6. Select an Approach. When it comes time to choose how to move forward, team members review the posted approaches and affix a sticky note to the one they like best. The project leader advises members of the team thusly: "Vote with your Post [sic]—not for an idea that's cool, but for an idea that's cool and buildable. If it's too far out there and it can't be built in a day, then I don't think we should vote for it." Due to the strict time constraint, there is some intervention taken on this step. A small group ("of self-appointed adults") narrows the list by taking the most common concerns and grouping them into the four categories noted above. Each concern-area team proceeds to address the overall problem from this more focused perspective.

7. Develop a Design Proposal. Due to the nature of the task and time constraints, this step is skipped.

8. Make a Model or Prototype. The shopping cart models and mock-ups are constructed from materials such as copper wire, PVC pipe, modular aluminum extrusion, foam core board, duct tape, modeling foam, construction lumber, electrical conduit tubing, and even part of an existing metal shopping cart.

9. Test and Evaluate the Design Using Specifications. Each concern-area team presents its model to the rest of the group at IDEO. The entire team provides constructive criticism on each design.

10. Refine the Design. The best elements from each of the four different concern-area models are combined to form the final solution. The cart is redesigned, rebuilt, and tested in an actual grocery store. The team gathers feedback from store employees. According to a cashier, the design "needs a little refining." The IDEO team leader notes that the cashier "gave us some good comments on how we could make this thing better."

Because this product isn't intended for actual production, the last two steps, **Create or Make the Solution** and **Communicate Processes and Results**, are not taken.

If you watch the video, you will notice that the IDEO team does not carry out the 12 steps of the engineering design process in order. After being presented with the problem (Step 1), the team jumps to Step 3 and begins researching and generating ideas. Steps 7, 11, and 12 are skipped entirely. As noted, the EDP is a tool—a template that helps engineers map and execute a specific approach to meet a specific challenge.

Incorporating Engineering: Why and Why Now?

Now that you have the foundation and terminology you need to acquaint yourself with the engineering landscape, we'll move forward with a look at why and how an infusion of engineering essentials can transform and unify math, science, and technology instruction.

The NGSS Connection

Scientific methods, various mathematical algorithms, and procedural steps in technology have been a staple in education for some time. Now, the Next Generation Science Standards (see www.nextgenscience.org) have increased the call to add basic knowledge of engineering and its processes to the standard curriculum. In the NGSS, engineering is presented as analogous to the major categories of science study.

When the National Research Council of the National Academy of Sciences (2012) published its *Framework for K–12 Science Education,* it identified the key scientific ideas and practices all students should learn by the end of high school. Both the framework and the Next Generation Science Standards that emerged from it have three dimensions: Disciplinary Core Ideas (DCI), Practices, and Crosscutting Concepts. Although there isn't a direct mention of engineering in the Crosscutting Concepts, it's a major presence in the other dimensions.

The DCI are grouped into four domains: (1) physical sciences; (2) life sciences; (3) the earth and space sciences; and (4) engineering, technology, and applications of science. For an overview of the engineering standards within the NGSS, please see the online supplement to this publication, available at www.ascd.org/ASCD/pdf/books/TruesdellSTEM.pdf. The practices place a strong emphasis on engineering. Indeed, their description posted on the

..... website, which states that they "describe behaviors that scientists engage in as they investigate and build models and theories about the natural world and the key set of engineering practices that engineers use as they design and build models and systems" (Achieve, Inc., 2013, para. 2), essentially gives engineering equal billing. The description goes on to say that bringing more engineering into the NGSS "will clarify for students the relevance of science, technology, engineering and mathematics (the four STEM fields) to everyday life" (para. 3). The thrust is that engineering makes science real.

This assertion is a game changer. Can you imagine if the Common Core State Standards (CCSS) for Mathematics stated that "science makes mathematics real"? Or if the Math CCSS included science as a domain or conceptual category? What we're seeing is a clear step toward breaking down the traditional barriers separating STEM content.

Given scheduling constraints, large class sizes, and the ever-increasing workload for teachers, it is no wonder that cross-curricular collaboration is relatively rare today. However, such teamwork can be transformative. Anyone who has taught mathematics, for example, knows that the concepts become more meaningful through application. What better way to teach algebra than to tie it into physics or chemistry formulas? In the NGSS, we find a bridge to the engineering silo. How the NGSS and their inclusion of engineering will play out remains to be seen, and we know these standards will have no formal impact on mathematics teachers, technology teachers, or even science teachers who live in states that chose not to adopt the standards. But now that the conversation has started, even these teachers will benefit from considering all that students stand to gain when the educational itinerary is expanded to include engineering.

The 21st Century Skills Connection

Certainly part of today's emphasis on STEM is related to the types of jobs projected to be available in the years ahead. However, engineering instruction is as much a vehicle for teaching life skills as it is preparation for potential careers.

The 21st century skills movement has stressed to educators and families that mastery of facts and figures does not sufficiently equip students for the world that they face. They must also be able to communicate, collaborate, make decisions, apply information, and articulate solutions for real-world problems—all skills that are required within the EDP. Let's look quickly at how the EDP's processes and the other components of the engineering toolbox line up with what are generally considered to be the "big three" of 21st century skills: (1) learning and innovation skills; (2) information, media, and technology skills; and (3) life and career skills (see, for example, www.p21.org).

Learning and innovation skills are based on critical thinking and problem solving. Given that the EDP starts with a problem to solve, its tie to this skill group is clear. "Complex communication" also falls under this umbrella. From brainstorming to team collaboration to presenting a solution to the client, communication is interwoven throughout the EDP. "Creativity and innovation," or "applied imagination and invention," are the final characteristics given for this skill (Trilling & Fadel, 2009). Many would say these terms could serve as a decent definition of engineering itself.

Information, media, and technology skills are all essential to engineering. Remember, Step 3 of the EDP requires students to "research and generate ideas." Internet research can be overwhelming, and students benefit from specific instruction aimed at teaching them to be discerning about online sources. Engineering projects can also help students hone practical technology skills as they learn how to use CAD software, create spreadsheets to present and examine data, create Gantt charts to plan their work (see an example at http://bit.ly/LfDjV0), and use presentation software to explain their results.

Life and career skills are a natural outgrowth of engineering education. Engineering taps into the interests of a wide variety of people. I used to tell my students that "whatever you might be interested in, there is a field of engineering to go along with it," often citing the example of my friend, a chemical engineer who worked in food making for the Sara Lee Corporation. Agricultural engineering might beckon to those who love to garden. Oceanic engineering might be a field for someone interested in dolphins. The list goes on and on.

Realistically, we know that only a certain number of students will go on to engineering careers, but that doesn't make pre-college study of engineering any less valuable. All students who engage with a curriculum that incorporates engineering will leave school with a higher level of

STEM literacy than they would otherwise have. Being able to reverse-engineer a product and examine its components will make them wiser consumers and more enlightened users of technology, no longer mystified by the function of cell phones or televisions. Understanding the environmental impact of putting certain substances into landfills will make them better stewards of the planet. These students will also leave school better citizens. Should tax dollars be spent to subsidize solar energy, electric cars, or fracking? Opinions will differ, but engineering education will have trained them to make their decisions based on reasoning and mathematical, scientific, and technological knowledge.

Engineering's Power to Engage

"Why do we have to learn this?" "When am I ever going to use this?" Anyone who has taught for any length of time has heard these questions and others like them. It is understandable that students want to understand the reason or purpose underlying their assignments, even if it is only to get a better grade. As adults, we feel the same way. Just think about a group of teachers in a professional development session that is totally irrelevant to their needs.

Good math, science, and technology teachers explain how classroom content connects to the world outside school. But incorporating engineering is a way to make these connections real. Creating solutions for real-life problems gives students a way to actually experience the significance of STEM content. Consider how common it is for young people to respond strongly to causes such as helping the poor and saving the planet. Engineering can provide a pathway for engaging authentically and creatively in these efforts. The related math, science, and technology take on a more personal meaning, and students are able to leverage their learning and their talents to benefit someone or something beyond themselves.

Service learning is an effective and practical way to connect STEM subject matter to engineering and authentic causes (see Lima & Oakes, 2006). I had the privilege of attending the 2012 American Society for Engineering Education's National Conference, where I heard William Oakes from Purdue University's Engineering Projects in Community Service (EPICS) program speak on the topic. EPICS deploys teams of undergraduates to design and build solutions for the engineering-related problems of local community service and education organizations. Founded at Purdue in 1995, EPICS soon spread to additional universities and then to the pre-college classroom. For example, at Frederick Douglass Academy in New York City, engineering and aquatic biology students in the EPICS High program have designed and built elevated outdoor aquaponics, irrigations, and hydroponics systems for community gardens around Harlem.

Although the National Youth Leadership Council (www.nylc.org) and Youth Service America (www.ysa.org) focus more on service learning in general than on engineering in particular, both offer a lot of resources, including conferences and webinars, that teachers can turn to for project ideas and ways to correlate these projects with course content in the STEM subjects.

To look at more global issues, a fantastic resource is the National Academy of Engineering's Grand Challenges—a list of 14 goals, ranging from making solar energy more affordable to providing access to clean water, that could be met though applied creativity and commitment. One advantage of presenting these challenges to students is that they don't know what "can't" be done. They naturally think outside of the box.

Planning and Delivering an Engineering-Infused Lesson

The time has come to address our overarching question: how does a teacher go about creating an engineering-infused science, technology, or math lesson that focuses students on finding solutions for real-world problems?

The Planning Stage

Planning begins with considering the standard or topic that is being taught. Clearly, some topics lend themselves to engineering proejcts more easily than others, and it's often a good idea to start off with a topic that allows engineering to be the foundation of the lesson before you tackle one where the connections are less obvious.

There is nothing wrong with using strong lessons that someone else has created. For sources, consider the websites www.teachengineering.org and www.tryengineering.org, both of which are stocked with such lessons. You could also explore prepared off-the-shelf curriculum. One of the best is the set of 20 units created by EiE, available for preview at www.eie.org/eie-curriculum. Each unit is modular and designed to be used in conjunction with an existing science curriculum. All lessons consist of a teacher guide with lesson plans, duplication masters, and assessments; a materials kit; and a storybook that expands the lesson's reach beyond STEM to language arts and social studies topics. Students are introduced to the topic through a story about a child, somewhere in the world, who has a practical problem to solve. With the guidance of a family friend or relative (who happens to be an engineer), the child solves the problem using the EDP. The catch is that students are not privy to the solution that the child in the story comes up with.; it's up to them to take the same information and solve the problem themselves.

Then there is the option of creating your own lessons. This allows you to address topics you are particularly excited about teaching and problems that really resonate with your students. You can design projects to draw upon their specific talents and interests, and you can incorporate various resources at your disposal, including video and reading material. Although designing a problem-focused, project-based engineering lesson is a complex undertaking, the general approach is relatively straightforward:

1. Identify your topic (e.g., the carbon cycle).

2. Identify the desired learning outcome, consulting standards as necessary (e.g., what students must come to understand about the carbon cycle).

3. Identify a real-world connection to the topic—some process or device that links this topic to students' experience (e.g., compost bins).

4. Consider how students could use the EDP to improve this process or device (e.g., students' task might be to design a better indoor compost bin). A word of caution: embarking on a project just because it seems like it would be fun or unique is never a good idea. You must take stock of how all aspects of the lesson will help your students meet the identified standards. At this point, too, you'll want to identify the kinds of materials required, consider a materials budget, and confirm that what you want to do is feasible.

5. Map out your schedule (e.g., how many days the lesson and the components of the lesson will require, including time estimates for the various steps in the EDP).

6. Determine—or refine your sense of—what information you will gather to evaluate student learning and when you will gather it. At this time, you must develop a set of rubrics that will be used for evaluation and will be shared with your students as you begin the project.

Let me pause to reiterate an essential point. Finding real-world applications will give you the most instructional "bang for your buck." Building a tower with marshmallows and toothpicks can be fun, but students will probably not remember why they did it. By contrast, building bridges with popsicle sticks that have to carry a load of five pounds is a lesson in the importance of structural design that will stick with students much longer.

It is also possible to present lessons focused more on engineering concepts than on design. You might, for example, want students to examine appropriate materials or tools, based on a specific set of properties. Testing the tensile strength of various plastics would fall into this category. Other lessons might look at the abstract effects of engineering products or processes. An example would be having students consider the impact that a move from gas-powered to electric cars might have on society, individuals, and the environment.

Whether you want students to design (or redesign) a product or focus on an engineering concept, I believe the best way to create an engineering project is to map it to the steps of the EDP:

1. Define the Problem
2. Brainstorm
3. Research and Generate Ideas
4. Identify Criteria and Specify Constraints
5. Explore Possibilities
6. Select an Approach
7. Develop a Design Proposal
8. Make a Model or Prototype
9. Test and Evaluate the Design Using Specifications
10. Refine the Design

11. Create or Make the Solution

12. Communicate Processes and Results

As noted, it's not always necessary or practical to have students work through all 12 steps, but they should be made aware of the process and see themselves following it. If a step is skipped, the reason for skipping it should be discussed.

Yes, designing this kind of lesson is ambitious. When you sit down to plan, it's easy to worry that students' engineering toolboxes aren't yet sufficiently stocked to produce a successful outcome. Don't worry about this. The most important thing in any engineering project is the process students follow. If the project is for students to design a better indoor compost bin, even if all of the students' designs fail miserably (which seldom happens), the lesson can still be a success. That's because by following the steps of the EDP, students will be reasoning, communicating, researching, making decisions, testing, evaluating, and learning course content. At the end of the project, they will also be reflecting on how they would improve on their designs in the next iteration. They will be building the skills they need to become better engineers.

It's always wise to approach creating an engineering lesson as if you were an engineer yourself. Yes, you design the lesson with the goal that students will be able to create something that will solve the problem you set, but there are no guarantees. The first iteration of your lesson may not work the way you wanted it to. Maybe the criteria or constraints were too restrictive. Perhaps the school custodian came in over winter break, saw the compost bins your students built, and threw them away before anyone could evaluate the rate of decomposition.

These outcomes are all opportunities for instruction. As teachers, we are models for our students. Facing and overcoming adversity are additional life skills they can learn from us. Engineers confront problems—it's what they do. A new problem means another opportunity. Once I embraced this attitude in my teaching and shared it with my students, I did not worry so much about presenting the perfect lesson or having all the answers. This didn't mean I stopped planning like crazy before executing a lesson; I just accepted that after the lesson, I could say to my students and myself, "What do you think should have been done differently?"

An Engineering Project, Illustrated

Being a Project Lead The Way teacher for 5 years gave me a great respect for PLTW's curriculum and the 12-step EDP it included. Through a National Science Foundation grant, I had engineering graduate students from the University of Cincinnati work in my class, and they confirmed that the processes we used in PLTW were authentic. Many of the strategies that I learned from PLTW I used to design new projects and units.

My favorite way to begin a project is to give students a design brief. As noted in our look at the *Nightline* shopping cart redesign challenge, a design brief corresponds to Step 4 of the EDP, **Identify Criteria and Specify Constraints**, and the information a design brief contains covers aspects of

Steps 1–3. In my design briefs, I generally identified myself as the client looking to hire one of the "companies" in my classroom (three- or five-person student groups) to respond to a challenge. I was careful not to present these companies as being in direct competition with one another. Instead, I stressed that in engineering there is no one right solution to a problem. Each solution has strengths and weaknesses. One may be more effective, while another one is cheaper to produce. The solution that receives the contract is not "better" but rather a better fit for the client's specific needs at that specific time.

If you were to start a science-based engineering project related to the carbon cycle in this way, you might generate a design brief that looks like this:

- Client Company: Pure Dirt Composters of America
- Target Consumer: Apartment dwellers or city dwellers with very little outdoor property
- Designer: _____. (Each team would have a "company" name.)
- The Problem: Landfill space is becoming an increasingly serious problem. Many environmentally conscious people compost to help alleviate the amount of material trucked off to landfills. Composting has the added benefit of producing rich soil. However, people who live in the city, and especially in apartments, may not have the outdoor space to compost.
- The Challenge: Your team of scientists and engineers has been hired to design an indoor composting

system. This system needs to be easy to use, turn kitchen waste into compost as quickly as possible, be inexpensive, and have little to no "side effects" (e.g., bugs, odor, mess).

- Criteria and Constraints: The materials for constructing the compost bin must cost less than $10. The maximum size is four cubic feet. The company will provide you with the materials to place within the bin to ensure consistency of comparison. This composition will mimic what a family would put in the bin. The maximum time for the material to decompose will be six months.

In presenting the design brief, take the opportunity to build up excitement in your students. In this case, have a little fun playing the role of CEO of Pure Dirt Composters of America (PDCA). And make sure that the teams of students choose names for their companies. It is important at the outset that this activity look different from the "group work" students have done before.

Although the approach I favor folds **Step 1—Define the Problem** into the design brief, it is a good idea to ask the students to restate the problem in their own words as an assignment to turn in. In large projects requiring multiple class sessions, having small but important pieces to turn in along the way helps keep students engaged, gives you insight into their progress, and allows you to provide targeted support as needed.

With the problem clear, the next step is to generate ideas for solutions. Before getting **Step 2—Brainstorm** under way, it's important to put students into their engineering teams. I am a firm believer in either three- or five-person teams; they are small enough for everyone to be actively involved, and the odd number eliminates split votes when team members are making decisions. You may assign teams or let students self-select into teams; there is value in both methods.

Here in Step 2, students must also know that there are definite rules to brainstorming. They should strive for quantity of ideas over quality of ideas, and there is to be NO criticism—positive or negative—when ideas are thrown out. It's obvious that telling someone that his or her idea is bad tends to shut everyone down, but praising contributions ("That's a fantastic idea!") can dissuade contributions just as easily. If the fantastic idea has already been given, why continue to offer more? Stress to that all ideas should be recorded.

Although ideas themselves should come from the students within each team, you may want to highlight how the design brief can point them toward factors they should consider. When brainstorming ideas for indoor compost bins, for example, students need to consider construction materials, bin shapes, where the bins will be stored, and so on.

Once all the ideas are on paper, it's time to refine the list. Students should review their team list, and then each individual should sketch the three or four ideas that he or she likes best. Explain that sketching is a way of communicating; they shouldn't be upset if they can't draw like Leonardo DaVinci. It is also important that they annotate their sketches. Small circles on a box could be airholes or polka dots; without an arrow leading to a label, how would one tell the difference?

After students have completed their sketches, ask them to answer the question, "Which is your best design?" with a clear explanation of why they think it is the best in light of the design brief's requirements and the lesson's learning objectives (e.g., what they know about the carbon cycle). This process holds every member of the team accountable for making an individual choice before the whole team begins to discuss which design they will choose collectively. It forces individual reflection.

As discussed, the steps of the EDP should be carried out in the order that suits the task and your students' reality. If they are unfamiliar with composting, for example, you might want to delay the brainstorming and have them complete **Step 3—Research and Generate Ideas** first. They might learn about composting by going to the library, through Internet searches, or by listening to a guest speaker from a local gardening group. Beware of allowing students to do too much research before they begin brainstorming, however. If they go online, pick an existing design for an indoor compost bin, and describe that in their brainstorming, they're not engaging in engineering.

What I have described thus far might take place over two or more class sessions. As noted, it is important to make careful estimates of the amount of time you intend to devote to each step in the process. Yes, you want to be flexible, but students need to know that engineers operate under constraints, and one of those is deadlines. Although a test score of 90 percent might translate into a classroom *A*, if an engineer presents a client with a prototype that is only 90 percent finished at the deadline, the company would probably lose the client's account, and the engineer might be out of a job. To keep students on track, I recommend requiring them to turn in something every day over the life of the project. It could be something as simple as a sheet of paper with answers to questions like these:

- What did our team accomplish today?
- What print or online resources did we use?
- What difficulties did we experience?
- What new/refined ideas did we come up with? (Provide a sketch if possible.)

I also like to ask my students to submit a weekly grade for themselves and each of their teammates, along with explanations of why these grades were earned. This idea came to me because of the main problem of group work: uneven distribution of the workload in which one person does everything but everyone gets credit. I find my students are usually very honest in their evaluations, even in the evaluation of their own work. I use these grades as a small percentage of the final project grade.

In terms of time, **Step 3—Research and Generate Ideas** and **Step 5—Explore Possibilities** are the steps where students tend to want to take the most time. What's more, students sometimes confuse these two steps, wanting to go on the Internet for both of them. Here's how to distinguish the two. *Step 3 looks at how things are.* What types of indoor composting bins currently exist? What do people who use them like about them or complain about? Online research can be very helpful here. *Step 5 considers how things could be.* Students should be looking inside their own heads. Given the criteria and constraints, what is possible? Could they significantly modify something that already exists, or should they create something completely original? A little research can be appropriate here, but it should focus on expanding ideas (e.g., what materials might help the students achieve what they want to do) and troubleshooting potential problems.

In between those two steps is **Step 4—Identify Criteria and Specify Constraints**. You, acting as the client, have a lot of latitude here. You could prescribe exactly what materials students are to use (here, it might be stating that PDCA has an old stockpile of plastic bins that would be perfect for this project). You could give students a resource list of materials that you will supply and allow them to choose what they think they will need. (This approach can be fun if you assign a "price" to each item and then give a maximum for students to "spend." There's value in students debating over trade-offs and figuring out how to do what they want to do within a budget.) Other options are to give a prescribed list of materials but allow teams one or two additional materials of their choice or to throw it open and allow teams to use whatever they want. If you choose the last option, which I have been hesitant to do, make sure that you set a price limit and require teams to produce evidence of how much they spent. Most projects can be done on a shoestring budget.

Some of the most creative things that I have seen involve items such as recycled milk jugs and duct tape.

In this example, because the aim is to compare compost bin designs, including how effectively they support decomposition, the variable of bin contents should be held steady. Specify the proportions of materials to add: how much newspaper, banana peels, coffee grounds, and "starter" compost. With all teams using the same input, the bin design will be a more significant factor in the bins' output. Of course, if the teams' designs are easy to construct, you might extend the lesson by asking them to use the "control" compost mix in one bin and their own recipe in a second, identical bin. And if you really wanted to be adventuresome, you might stipulate that the second bin use a different composting technique, such as vermicomposting or bokashi.

Up to this point, the compost bins exist only on paper and in the minds of the students. Give students a deadline by which they must complete **Step 6—Select an Approach**. A decision matrix is a key tool for this step, as it helps avoid the hurt feelings that can arise when people take ownership of an idea and get upset when others don't share their enthusiasm. With a decision matrix, each team picks the characteristics of the design that they think are most important. For an indoor compost bin, those might include "cost" and "avoids bad smells." Each characteristic is given a weight and the bins are "graded" against these criteria.

In an engineering course, **Step 7—Develop a Design Proposal** would involve CAD software and address additional specifics on materials; this is probably not necessary for a project like the compost bin challenge, especially if you are keeping the lesson focus on the carbon cycle or other ways to break down food waste. However, it is a good idea for teams to create a more detailed annotated sketch before they begin building. This step will allow them to confirm what materials they need and how those materials will be put together. It underscores the importance of planning.

These plans will be implemented in **Step 8—Make a Model or Prototype**. Some people confuse or misuse the terms *model* and *prototype,* so be specific with students on which they will be making (for the compost bin, a prototype) and what the difference between the terms is. Unlike a prototype, a model does not need to work and can be proportionally smaller than the actual final product is intended to be. A good example of this is found in the automotive industry, where manufacturers create a clay model of a new car design, usually about one-quarter of the actual size, to test the aerodynamics in a wind tunnel. However, before the car goes into production, a full working prototype will drive laps around a track.

Here again, deadlines are crucial. Most compost bin designs are not incredibly intricate, so students should be able to create and fill them relatively quickly.

In general, **Step 9—Test and Evaluate the Design Using Specifications** can take a long time or not much time at all. For this particular project, which involves decomposition, Step 9 would require several weeks at least, although active engagement would be minimal, limited to daily or weekly temperature readings or interventions to deal with issues such as seepage or fruit flies. All of these "check-ins" need to be documented. The team should have a notebook or a reporting sheet to capture whatever information you deem important. Make sure to encourage sketching—identifying where a bin was leaking or what the mold on the outside looked like will be helpful when it comes time to evaluate or refine the design. It is also important to note when the decomposers, the organisms that feed on dead organic matter, appear. Sketching these helpful creatures to identify the species and the actions they perform would definitely be beneficial, not only to the project itself but also as a way to pull in other fields of science.

It will be up to you, the teacher, to determine when to conclude the testing period. In this project, it's best to take a peek and confirm that at least of couple of the bins' contents have decomposed significantly. Factors that would affect the rate of decomposition include the compost recipe used, the quality of the bin designs, the number of times students turn the compost, and even the temperature of the classroom.

The decision to include **Step 10—Refine the Design** and **Step 11—Create or Make the Solution** often depends on how much time a teacher can afford to devote to a project. They are the two steps in the EDP that teachers tend to skip due to time constraints. At least once a year, it is helpful to create a simpler project that allows time for these steps. In my work for the University of Cincinnati, I saw a lesson set up in such way that students completed the entire EDP in a single class period. This lesson ("The Catch," created by Deon Edwards) is available online at http://bit.ly/1loCpGz.

Every engineering project must always include **Step 12—Communicate Processes and Results**, as it is the key to solidifying students' understanding of the STEM concepts addressed in the lesson. This step is generally conducted as a series of 5- to 10-minute presentations, during which each team of students reflects on and synthesizes what happened throughout the entire process. If there wasn't enough time to do Step 10, teams can comment on how they might refine the design for a second iteration. If you have the technology at your disposal, students might record these presentations on video and post them to your school's Learning Management System (e.g., Blackboard, Moodle).

What about assessment? In this overview, there have been references to various products you should expect from your students and their teams. These products, including brainstorming documentation, design proposal sketches, reflections, daily logs, decision matrices, documentation of research and discussions, results from product testing, final presentations, and the actual compost bins, provide ample material for both formative and summative assessment of the students' learning. It is also important that, at the beginning of the project, students be provided with rubrics for the major components, such as the compost bins and the final presentation. In the Encore section of this publication, you will find a template for gathering all the intermediate assessments to document students' engagement in the EDP and an illustration of the kind of rubric that could guide both student work and its evaluation.

Aside from the rubrics you select or develop, there are curriculum requirements and standards to be met. A science-related project like the compost bin challenge should be considered in light of the NGSS. For example, within the standard focused on analyzing and interpreting data on resource availability within ecosystems (MS-LS2-1), one of the Crosscutting Concepts that deals with Cause and Effect states, "Cause and effect relationships may be used to predict phenomena in natural or designed systems" (NGSS Lead States, 2013). In filling the compost bins with layers of "greens" and "browns," students might reflect on how their designed system mirrors the natural compost creation in a forest and speculate as to why this layering method should work in both scenarios. You might even wish to create a "test" compost bin of your own that does not use the natural system's method, get predictions on what will happen, and eventually discuss why your bin generates different outcomes.

The compost bin project is very much focused on the "S" in STEM, but a project focused on the "T" or the "M" would follow a similar pattern. Technology or math could also be incorporated into this example. Students might examine the types of plastics that teams are using and the possible impact of compounds such as BPA. They might determine which materials are permissible in the compost recipe if soil is to be considered "organic." They might use the Vernier sensors to measure temperature, pH, and oxygen gas levels. Even taking these measurements with more traditional tools, such as liquid-in-glass thermometers, is using technology.

Measurements, once obtained, are a rich resource for math classes. Graphing and analyzing this kind of real-world data is much more meaningful than working with the numbers provided by a textbook publisher. Whether the shape of the compost bins was a factor would be a good question for Geometry teachers to ask. They might also direct students to look at the relationship between each bin's effectiveness and its surface area and volume.

Is a unit like this one really worth the time that it takes? First, consider that its cross-curricular benefits would justify dividing the work among several subject-area teachers, so the time it takes doesn't necessarily have to be your time alone. That said, the simple answer is yes, this kind of unit *is* worth it. Engineering projects that ask students to design solutions for real-world problems, when developed thoughtfully and implemented according to the EDP, pay dividends that last. The compost bin design project we have looked at was inspired by a lesson done by Jeanette McNally (2012), a participant in the University of Dayton's Research Experiences for Teachers program. Her unit, called "Umm, You're Eating My Trash," began with compostable snack wrappers, and it was tied into composting, landfills, engineering, and Earth's four spheres (geosphere, hydrosphere, biosphere, and atmosphere). She used it as part of a discussion of a lesson on the carbon cycle, which is one of the biogeochemical cycles in science. Later, when her class began studying the hydrologic cycle (water cycle), one of her students said,

"Hey, doesn't that tie into the composting stuff that we did?" What greater satisfaction can a teacher have than when the students connect the dots themselves?

Conclusion

In looking back over this whirlwind tour of engineering for the STEM classroom, there are some points of the trip that need to be emphasized.

First, engineering requires a group, not a lone traveler. Even though solitary inventors do exist, the use of teams will maximize the engineering experience in the classroom.

Second, engineers must all speak the same language. Communication is crucial within the team, particularly when members are engaged in processes such as brainstorming or are reporting their accomplishments and what they have learned. When working with your students, be sure to use the vocabulary of the EDP and teach it as you would teach the content-area vocabulary.

Finally, engineering isn't just a trip, it is an adventure. Whatever problem is presented to the class, it must have an open-ended result. Not only is there never one "right" answer, but there must always be many answers. If everyone is testing exactly the same thing in the same way, what you are doing is not engineering.

When we go on vacation, my family plans stops along the way at our favorite destinations. My husband truly believes that half the fun is getting there. In engineering, more than half the learning is getting there. Engineering, by nature, is a process, not a product. Even if the product fails, the team of student engineers can succeed. The learning that can result from mistakes will outlive even the best-designed indoor compost bin.

I have been where you are now. During my 25 years as a computer science and mathematics teacher, I often struggled to find real-world ways for my students to apply the skills and concepts that I was working so hard to teach them. Then, when my position as a computer science teacher was cut and my options were to teach engineering or find another school, I journeyed into engineering's unfamiliar territory and found the approach I'd been seeking for so long.

Whether you teach science, technology, or mathematics, you now have the essentials of this wonderful world of engineering. I wish you *bon voyage* on the first of many adventures. Don't forget to write!

To give your feedback on this publication and be entered into a drawing for a free ASCD Arias e-book, click here or type in this web location: **www.ascd.org/ariasfeedback**

ASCD | arias™

ENCORE

ASSESSING AND DOCUMENTING THE ENGINEERING DESIGN PROCESS

In lessons that incorporate engineering, assessment takes on a different look and feel. Because the focus is the process and not the product, evaluating students on the steps that they follow to arrive at a final product is crucial. The final product is important as an outgrowth of the quality of work students have done.

This publication's look at a *sample* engineering-infused lesson included a number of ideas for intermediate assignments. Here, they are presented in a more generic format that could apply to any project. In my experience, these types of assignments are necessary on a daily or weekly basis. The resulting documents can also be compiled into a portfolio and used for the completion of Step 12 of the EDP.

○ STEP 1—Define a Problem. Having reviewed the design brief for this project, state the problem in your own words.

○ STEP 2—Brainstorm. Before making any final decisions with your team members, you need to consider all ideas. Provide the teacher with documentation of your team's brainstorming. Provide a minimum of three design concept sketches using one sheet of graph paper. Fold the sheet in half so that you will have two areas on the front and two on the back. Make sure to use most of the space for your sketches. Sketches must be annotated to receive the full number of points. On a separate piece of paper, write one to two paragraphs explaining which design is the best. Explain the reasons for your choice.

⭕ STEPS 3–8: Research and Generate Ideas; Identify Criteria and Specify Constraints; Explore Possibilities; Select an Approach; Develop a Design Proposal; Make a Model or Prototype

Daily Research & Design Journal. Each day, you will record and submit a document with the following information. You may use bulleted lists. Complete sentences, correct grammar, and clarity are required.

- Today we completed the following tasks/activities:
- Today we used the following resources (list those from the Internet, books, or articles):
- Problems that we encountered today (leave this blank if there were no problems):
- Sketches (include as needed to document developments or clarify progress):

Weekly Team Evaluation. List the names of your team members and give them a grade (1–5) based on their contribution to the team's work. Using complete sentences and correct grammar, explain why each earned this grade.

⭕ STEP 6—Select and Approach (using a decision matrix or Pugh chart). Turn in the chart that your team creates.

One person on the team needs to include a paragraph explaining the approach you have chosen and how you arrived at that decision. Make sure to include whether the decision was unanimous or required a compromise.

⬭ STEP 7—Develop a Design Proposal. Submit a highly detailed, annotated sketch of your team's chosen design.

⬭ STEP 8—Make a Model or Prototype. Create and submit an actual product. If a prototype, it should be full-scale and do what it is designed to do.

⬭ STEP 9—Test and Evaluate the Design Using Specifications. Depending on the project, this step can vary greatly. For the compost bins, it could be a series of standard, data-gathering questions such as the following:

1. Today is Day ___ of the compost bin being full.
2. The temperature of the bin is ____.
3. List any problems with the bin. Include sketches if necessary.
4. Report on any decomposers evident in the bin; include sketches.
5. (Final Day Only) Weigh the bin contents and sort out decomposed and nondecomposed material. The weight of the decomposed material is ____. This is ___ percent of the final material weight. The weight of the nondecomposed material is ____. This is ___ percent of the final material weight.

◯ STEP 12— Communicate Processes and Results.
Produce a presentation (using a poster or PowerPoint,
Keynote, Prezi, or other presentation software). The entire
team will participate in the presentation, dividing up
topics to be covered equally.

RUBRICS

Even though the intermediate assignments are valuable,
the bigger picture can't be lost. Giving students a rubric
covering the major components of the project clarifies the
long-term learning goals.

Generic rubrics are difficult to create, as a good rubric will
focus on the specifics of the project. I hope the example
that follows, a sample rubric for the compost bin project,
provides adequate illustration to support its adaptation to
other types of projects.

A Sample Rubric for the Compost Bin Project

Project Component	4 – Expert	3 – Competent	2 – Beginner	1 – Novice
Evidence of research	The team documents research on appropriate compost bin design/materials and compost ingredients; all references are credible.	The team documents research on appropriate compost bin design/materials and compost ingredients; most references are credible.	The team documents research on appropriate compost bin design/materials and compost ingredients; there is least one credible reference.	The team provides no documentation of research.
Evidence of following the engineering design process (EDP) *Note:* This is where a portfolio comes into play.	The team documents all steps of the EDP, including evidence of multiple design considerations.	The team documents most steps of the EDP, including evidence of two or more design considerations.	The team documents some steps of the EDP.	The team provides no documentation of following the EDP.

Compost bin design	The bin meets size specifications ($\leq$ 4 cubic feet) and allows access for accurate measurement; bin allows adequate room for turning the compost; bin is portable; documentation shows materials cost less than $10.	The bin meets size specifications ($\leq$ 4 cubic feet) and allows access for accurate measurement; bin allows adequate room for turning the compost; bin is portable; documentation shows materials cost less than $10.	The bin meets size specifications ($\leq$ 4 cubic feet) and allows access for accurate measurement; bin does not allow adequate room for turning the compost; bin is not portable; materials appear to cost less than $10, but documentation is not provided.	The bin does not allow access for accurate measurement or is larger than 4 cubic feet; bin does not allow adequate room for turning the compost; bin is not portable; materials cost more than $10, or the cost is not documented.

(continued)

A Sample Rubric for the Compost Bin Project *(continued)*

Project Component	4 — Expert	3 — Competent	2 — Beginner	1 — Novice
Data collection with Vernier probes	The team uses multiple probes; provides evidence of data; data collected is relevant.	The team uses multiple probes; data collected is relevant.	The team uses at least one probe; data collected is not always relevant.	The team supplies no evidence of data collection and/or the data collected is not relevant.
Compost bin functionality	There is evidence that the compost bin is working, the compost is odorless, and the ratio of output volume to input volume is high.	There is evidence that the compost bin is working, and the compost is relatively odorless.	There is evidence that the compost bin is working.	The compost bin is not working; there is no compost.

Team presentation	On average, team members exhibit outstanding use of voice, posture, eye contact, gestures, pacing; presentation shows expert incorporation of pictures, graphs, or computer models; content is interesting and vivid; there is excellent use of original photos and graphics.	On average, team members exhibit satisfactory use of voice, posture, eye contact, gestures, and pacing; presentation shows good incorporation of pictures, graphs, or computer models; there is good use of original photos and graphics.	On average, team members exhibit marginal use of voice, posture, eye contact, gestures, pacing; pictures, graphs, or computer models are present but not used very effectively; there is very little use of original photos and graphics.	On average, team members exhibit poor use of voice, posture, eye contact, gestures, pacing, or seem tense or unprepared; pictures, graphs, or computer models are absent or not used effectively; there is no use of original photos and graphics.

(continued)

A Sample Rubric for the Compost Bin Project *(continued)*

Project Component	4 – Expert	3 – Competent	2 – Beginner	1 – Novice
Evidence of teamwork with individual responsibility	Documentation shows the design was chosen as a team through thoughtful deliberation; all team members are able to explain to the class what they did throughout the project.	Documentation shows the design was chosen as a team after some deliberation; most team members are able to explain to the class what they did throughout the project.	Documentation shows the design was chosen as a team after a little discussion; some team members are able to explain to the class what they did throughout the project.	Documentation shows that the design was chosen as a team hastily and without much discussion, or documentation is insufficient to evaluate the team's decision making related to design choice; only one team member is able to explain to the class what he or she did throughout the project.

The

References

ABC News [producer]. (9 February 1999). The deep dive: One company's secret weapon for innovation. *Nightline.* Available: http://www.youtube.com/watch?v=2Dtrkrz0yoU

Achieve, Inc. (2013). *Next Generation Science Standards: For states, by states* [website]. Available: http://www.nextgenscience.org/

American Heritage Dictionary. (2009). Engineering. Retrieved January 9, 2014, from http://www.thefreedictionary.com/engineering

American Society for Engineering Education. (2009, October). *Engineering, Go For It.* Available: http://students.egfi-k12.org/eGFI-Engineering-Go-For-It-Magazine.pdf

Cunningham, C. M. (2009, Fall). Engineering is Elementary. *The Bridge, 39*(3), 11–17.

Engineering is Elementary. (2012). The engineering design process. Available: http://www.eie.org/overview/engineering-design-process

Lima, M., & Oakes, W. C. (2006). *Service-learning: Engineering in your community.* New York: Oxford University Press USA.

McNally, J. (2012). Umm, you're eating my trash [lesson plan]. Dayton Regional STEM Center. Retrieved January 9, 2014, from http://daytonregionalstemcenter.org/curriculum/umm-youre-eating-my-trash

National Research Council of the National Academy of Sciences. (2012). *A framework for K–12 science education: Practices, crosscutting concepts, and core ideas.* Washington, DC: The National Academies Press.

NGSS Lead States. (2013). *Next Generation Science Standards: For states, by states.* Washington, DC: The National Academies Press.

Project Lead The Way. (2007). *Introduction to engineering design 2007 – 2008.* Indianapolis, IN: Project Lead The Way.

Trilling, B., & Fadel, C. (2009). *21st century skills: Learning for life in our times.* San Francisco: Jossey-Bass.

Related Resources

At the time of publication, the following ASCD resources were available (ASCD stock numbers appear in parentheses). For up-to-date information about ASCD resources, go to www.ascd.org. You can search the complete archives of *Educational Leadership* at http://www.ascd.org/el.

ASCD EDge® Groups
Exchange ideas and connect with other educators interested in STEM topics on the social networking site ASCD EDge at http://ascdedge.ascd.org.

Print Products
Concept-Rich Mathematics Instruction: Building a Strong Foundation for Reasoning and Problem Solving by Meir Ben-Hur (#106008)

Engaging Minds in Science and Math Classrooms: The Surprising Power of Joy by Eric Brunsell and Michelle A. Fleming (#113023)

Inference: Teaching Students to Develop Hypotheses, Evaluate Evidence, and Draw Logical Conclusions (A Strategic Teacher PLC Guide) by Harvey F. Silver, R. Thomas Dewing, and Matthew J. Perini (#112027)

Learning to Love Math: Teaching Strategies That Change Student Attitudes and Get Results by Judy Willis (#108073)

Succeeding with Inquiry in Science and Math Classrooms by Jeff C. Marshall (#113008)

Video
The Innovators: STEM Your School (DVD) (#613042)

For more information: send e-mail to member@ascd.org; call 1-800-933-2723 or 703-578-9600, press 2; send a fax to 703-575-5400; or write to Information Services, ASCD, 1703 N. Beauregard St., Alexandria, VA 22311-1714 USA.

About the Author

Pamela Truesdell is a member of the Resource Team for the Cincinnati Engineering Enhanced Mathematics and Science Program (CEEMS) at the University of Cincinnati, where she coaches teachers as they implement challenge-based learning through engineering-infused lessons. A teacher for 30 years in the Cincinnati Public Schools, she taught history, mathematics, computer science, and engineering. She served as the chairperson for the Engineering Department at Western Hills Engineering High School and was part of the committee that helped the school receive National Certification from Project Lead The Way Engineering in 2010. In 2011, Truesdell was selected for the Albert Einstein Distinguished Educator Fellowship Program, and she served two years at the National Science Foundation working on the Research Experiences for Teachers in Engineering program. You may reach her at pamela.truesdell@gmail.com.

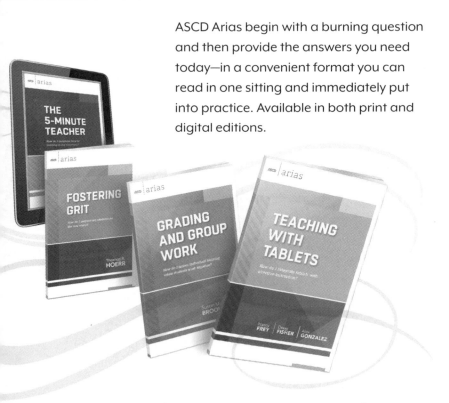